Liaison Functions 2022

Held at the 2022 AIChE Spring Meeting and 18th Global Congress on Process Safety

San Antonio, Texas, USA
10 – 14 April 2022

ISBN: 978-1-7138-6831-6

Printed from e-media with permission by:

Curran Associates, Inc.
57 Morehouse Lane
Red Hook, NY 12571

Some format issues inherent in the e-media version may also appear in this print version.

Printed with permission by Curran Associates, Inc. (2023)

For permission requests, please contact AIChE
at the address below.

AIChE
120 Wall Street, FL 23
New York, NY 10005-4020

Phone: (800) 242-4363
Fax: (203) 775-5177

www.aiche.org

Additional copies of this publication are available from:

Curran Associates, Inc.
57 Morehouse Lane
Red Hook, NY 12571 USA
Phone: 845-758-0400
Fax: 845-758-2633
Email: curran@proceedings.com
Web: www.proceedings.com

TABLE OF CONTENTS

641154 The Chemical Plant of the Future: Progressing Toward Digitization and Sustainability

Abstract:

Two things are driving the plant of the future and are indeed significant challenges we see in our future. The first is "The Age of Digitalization" and the second is the "Age of Sustainability." A discussion of the transformation of our facilities and our industry beyond 2050 would be incomplete without a deeper look at what a digital, sustainable facility of the future will look like.

640485 “The Role of Carbon Capture in Meeting Net-Zero Carbon Goals”

Abstract:

President Biden has laid out a bold and ambitious goal of achieving net-zero carbon emissions in the U.S. by 2050. The pathway to that target includes cutting total greenhouse gas emissions in half by 2030 and eliminating them entirely from the Nation’s electricity sector by 2035. Investment in technology research, design, development, and deployment (RDD&D) will be required to achieve the president’s objectives, including investments in both carbon capture at point sources in addition to carbon dioxide removal approaches that target the accumulated pool of carbon in the atmosphere. Both will be required to achieve net-zero carbon emissions in time and they will require increased deployment in order to move down the cost curve. These efforts combined with effective policy will make these approaches economically viable.

These approaches are critical and they must be deployed in parallel. Deployment of these technologies at the scale required will necessitate the use of resources including land, water, and in some cases, low-carbon energy, while ensuring the secure and reliable storage of carbon dioxide (CO_2) on a timescale that impacts climate. Therefore, CCS and CDR deployment must be implemented strategically in terms of regional goals and requirements.

The Office of Fossil Energy and Carbon Management will play an important role in the transition to net-zero carbon emissions by reducing the environmental impacts of fossil energy production and use – and helping decarbonize other hard-to abate sectors – through investments in technology solutions including CCS, direct air capture, and the deployment of carbon capture technologies to produce low-carbon products and fuel, including hydrogen.

640947 Process Safety Culture Development: Beyond the Metrics

Abstract:

Digitalization has done much to enable data transparency over the last decades. Today's tools provide data transparency necessary to move to more proactive management programs. This talk focuses on aspects of process safety culture development that have proven to be impactful in the manufacturing environment.

641610 Navigating the Energy Transition

Abstract:

There is an emerging global consensus that steps must be taken to mitigate the impact of anthropogenic climate change. The Paris Agreement on climate change has been ratified by 192 countries (all countries except Iran, Libya, Eritrea and Yemen) and the signatories have pledged to make changes to their patterns of energy and land use that achieve "carbon neutrality" or net-zero emissions of greenhouse gases (GHG) by approximately mid-century. In these countries, energy ministries, energy companies and utilities are evaluating alternative fuels and power sources that can deliver the heat and power required for a modern economy with reduced GHG emissions. While technically proven low-emissions alternatives exist for almost every application, most of these alternatives cost substantially more than the fuels or energy sources they replace. Consequently, most countries will use a combination of regulations, taxes and subsidies to distort the energy market in favor of the lower-emissions alternatives. Achieving a net-zero GHG economy while meeting the energy demand and economic growth needs of the world's population requires a substantial transformation of the energy industry. Scenario analysis can be used to develop rational models of the pace at which the energy transition could occur, the resulting cost impact on society and the effect on global climate. Current rates of decarbonization of the energy supply are not consistent with meeting the goals of the Paris Agreement, as can clearly be seen by tracking important energy transition marker variables.

NOTES

NOTES

NOTES

NOTES

NOTES

NOTES

AUTHOR INDEX